Fluid-Structure Interaction of Symmetrical and Cambered Spring-Mounted Wings

Using Various Spring Preloads and Pivot Point Locations

Fluid-Structure Interaction of Symmetrical and Cambered Spring-Mounted Wings

Using Various Spring Preloads and Pivot Point Locations

Simon Fels et al.

ELIVA PRESS

ISBN: 978-99949-8-230-1

Cover Design: Eliva Press
Cover Image: Freepik Premium
Email: info@elivapress.com
Website: www.elivapress.com

This Eliva Press imprint is published by the registered company Eliva Press Global Ltd. part of Eliva Press S.R.L. Publishing Group

The registered company address is: Pope Hennessy Street Level 2, Hennessy Tower Port Louis, Mauritius
Eliva Press S.R.L. Publishing Group address is: Bulevardul Moscova 21, Chisinau, Moldova, Europe

ABSTRACT

The fluid–structure interaction of a pivoting rigid wing connected to a spring and subjected to freestream airflow in a wind tunnel is presented. Fluid–structure interactions can, on the one hand, lead to undesirable aerodynamic behaviour or, in extreme cases, to structural failure. On the other hand, improved aerodynamic performance can be achieved if a controlled application within certain limitations is provided. One application is the reduction of drag of road vehicles at higher speeds on a straight, while maintaining downforce at lower speeds during cornering. Conversely, another application concerns increased downforce at higher windspeeds, enhancing vehicle stability. In our wind tunnel experiments, the angle of incidence of the spring-mounted wing is either increased or decreased depending on the pivot point location and spring torque. Starting from a specified initial angle, the aerodynamic forces overcome a pre-set spring preload at incrementally increased freestream velocity. Reynolds numbers at a range of $Re = 3 \times 10^4$ up to $Re = 1.37 \times 10^5$ are considered. The application of a symmetrical NACA 0012 and a cambered NACA 6412 airfoil are tested in the wind tunnel and compared. For both airfoils mounted ahead of the aerodynamic centre, stable results were achieved for angles above 15° and below 12° for the symmetrical airfoil, and above 25° and between 10° and −2° for the cambered airfoil. Unsteady motions were observed around the stall region for both airfoils with all spring torque settings and also below −2° degrees for the cambered airfoil. Stable results were also found out-side of the stall region when both airfoils were mounted behind the aerodynamic centre, although the velocity ranges were much smaller and highly dependent on the pivot point location. An analysis is reported concerning how changing the spring torque settings at each pivot point location effects performance. The differences in performance between the symmetrical and cambered profiles are then presented. The oscillation in the near-stall region is investigated in more detail for the case of the symmetrical wing. It is identified as a periodic, self-sustaining oscillation that has dimensionless reduced frequencies in the range of 0.14 to 0.22. Finally, an evaluation of the systems' effects was conducted with conclusions, future improvements, and potential applications.

Table of content

Table of figures

1. INTRODUCTION

Fluid–structure interaction describes the relationship between fluid flow and motion or deformation of a structure [1,2]. Fluid–structure interaction uses multi-physics coupling between laws describing fluid dynamics and structural mechanics [3–5]. The phenom-enon is characterised by unstable, oscillatory, or stable interactions between deformable or moving structures and fluid flow [6]. Stresses exerted on structures lead to strains within that structure, which if sufficiently high can cause the structure to deform. The defor-mation endured depends on the material properties and the configuration of the structure as well as the pressure exerted by the fluid [3,7].

Aeroelasticity describes the interaction of flexible structures with a surrounding airflow [8–10]. Aircraft wings are an example of a flexible structure [11,12], which are prone to the aeroelastic effects of divergence, control reversal, and flutter [13,14]. Most flight conditions have aerodynamic and elastic loads that converge into a state of equi-librium (static aeroelasticity). When the loop becomes unstable, the deformations increase without or with oscillations (dynamic aeroelasticity), which can both lead to structural failures. Although understood, research is still highly active in this area, aiming to in-crease the accuracy in predictions and efficiency of such tools [15–17]. Linear models in the calculation of flutter have been proposed by Stancui [18]. The flow-induced vibration in the deep stall has been investigated by Sidlof [19]. The effect of flutter on the NACA 0012 has been investigated experimentally by Rivera [20] and also numerically by Khatir [21]. The NACA 0012 was also used to investigate flutter in the transitional range by Goyaniuk [22].

Morphing of airfoils affects the aerodynamic performance by changing its profile, resulting in revised aerodynamic forces and flow fields [23–27]. This type of interaction can lead to the improved performance of airfoils for their relevant application by a means of an active nature. The work conducted in this paper is based upon that of a passive nature using a rigid fixed surface of an airfoil that does not deform. The rigid airfoil is attached to a spring that leads to a fluid–structure interaction which can result in improved performance for a prescribed condition. For example, a race car requires a relatively large amount of downforce when cornering; however, this increased level of downforce will cause large amounts of drag on the straights [28,29]. Therefore, by using a method of a passive nature, the amount of drag on the straights can be decreased above a certain vehicle speed, while maintaining the level of downforce below a certain speed on the corners, thereby improving efficiency and decreasing lap times [30,31]. Conversely, the spring-wing system could be set up so that increasing the incidence leads to more downforce and drag, which would aid stability at higher speeds [32]. Other potential applications of this work could include the fluid–structure interaction of hydrofoils to improve performance [33,34], as well as managing forces on sails used in shipping or wind turbines [35,36] in order to prevent damage in harsh conditions. The work could be applied to the protection of a workforce by using fences that moderate flow while maintaining levels of ventilation in harsh conditions [37]. Certain micro air vehicle and unmanned air vehicle applications could also potentially benefit from passive aerody-namic systems [38–40]. Wind tunnels enable the performance of these interactions to be easily identified and analysed for a wide variety of configurations over a relatively short timescale, allowing for the optimal performance to be found for a particular application of the airfoil or aerodynamic surface [41].

The system is set up similar to the work of Poirel et al. [42], except we omit the effect of heave in our approach. The proof of concept and validation of our system has been shown by Knight et al. [43] for the symmetrical NACA 0012 mounted at 10% chord. The aim of the work reported here is to further analyse the fluid–structure interaction of the symmetrical NACA 0012 with various pivot point locations and to compare this with the cambered NACA 6412. The effects that windspeed, pivot point locations, and spring torque have upon the system's structure and the resulting flow characteristics on both airfoils are reported. The overarching aim is to ensure reduced drag at higher windspeeds when using a pivot point location ahead of the aerodynamic centre. Conversely, if in-creased stability is desired, an increased downforce at higher windspeeds can be realised by using a pivot point location behind the aerodynamic centre. In this work for increased stability, we only report the post-stall region whereby stable results are found.

Testing was carried out within the wind tunnel of the University of Portsmouth by varying the airflow velocity and spring torque for each airfoil at various pivot point lo-cations and measuring the change in angle of incidence of the airfoil. This enabled the analysis of how changing the spring torque settings at each pivot point location for both airfoils effects the performance, allowing for the comparison of how changing the pivot point location and spring torque settings have differing effects for the two different airfoils. Finally, conclusions and the evaluation of the system effects were conducted in the performance with future improvements and considerations discussed.

2. METHODOLOGY

The present study builds upon an aeroelastic system, in which a ridged wing self-adjusts its angle of incidence to the airflow velocity using an elastic element. The system was also presented and verified at low angles in a previous study by Knight et al. [43]. The general system uses a wing mounted upon a rod with only a rotational degree of freedom in pitching direction. The elastic element of the system is a spiral spring, which has an adjustable preload angle (θ). It is attached to the rod and casing. The force required to overcome the springs preload, causing the airfoil to rotate, occurs mainly due to the aerodynamic pitching moment, although weight is also taken into account.

The basic system is extended to allow for experiments in which the wing is mounted further back from the leading edge in order to investigate the effect of the position of the pivot point location on the behaviour of the system. Furthermore, the targeted reversed scenario, in which the wing is expected to increase its angle of incidence, can be tested by utilising pivot point locations after the aerodynamic centre and negative spring preload angles. In addition to the use of a symmetrical NACA 0012 profile, a second wing with the NACA 6412 profile is also used to investigate the application of a cambered airfoil profile. We acknowledge that many different profiles of varying camber and thicknesses are used in motorsport; however, we choose NACA 0012 and NACA 6412 as they are both very common in research literature, enabling comparisons with other research findings [44,45]. The aeroelastic system and its features are shown as a schematic in Figure 1.

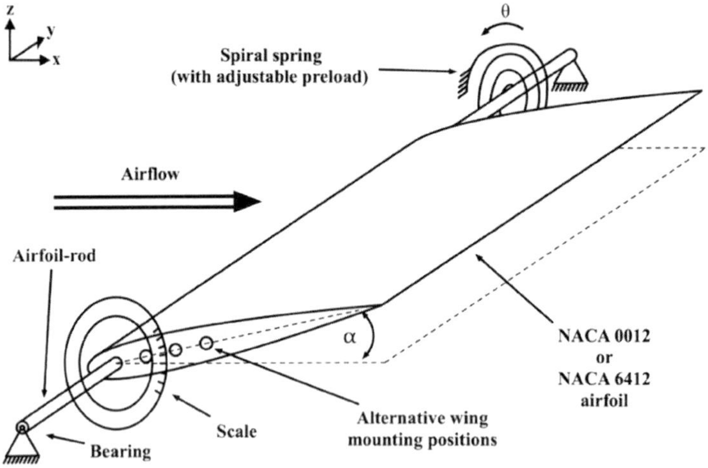

Figure 1. *Schematic of the aeroelastic system.*

2.1. System Characteristics

The system characteristics of the proposed aeroelastic system are of a mechanical and aerodynamic nature. When considering a static equilibrium of forces, the system is mainly determined by the spiral springs' characteristics, aerodynamic forces, and the weight of the airfoil. For the purpose of this work, inertial, frictional, and time dependency effects are ignored. The total moment is produced from the weight of the wing (τ_g), the springs torque (τ_s) and the airflow (τ_a), with the rotational direction for the leading edge moving upwards taken as positive (Equation (1)).

$$\tau_g + \tau_s + \tau_a = 0 \qquad (1)$$

Although the wings used throughout this work are lightweight, the gravitational force on the wing will affect the system. The wings' weight (m_w) will remain constant throughout this work with the moment of τ_g varying with the angle of incidence (α), which is calculated using trigonometry. The acceleration due to gravity is g and the distance from the leading edge to the centre of gravity (CoG) is x_{CoG}. The distance from the leading edge to the mounting axis will be varied and is expressed as x_{MPi}. Thence, τ_g is given in Equation (2) for the four pivot point locations used in this work.

$$\tau_g(\alpha) = \cos(\alpha)\,(x_{CoG} - x_{MPi})\; m_w\, g \quad for\ i = 1, 2, 3, 4$$

where

$$x_{MPi} = 10\ \text{mm} + 17.5\ \text{mm} * (i - 1). \qquad (2)$$

Accordingly, the gravitational moment will become negative for a pivot point location with a distance greater than the distance of the centre of gravity from the leading edge.

The moment for the springs torque (τ_s) is defined assuming the spring is used only in the linear-elastic range. This allows Hooke's Law to be applied with the tensioning of the spring at a certain angle, acting as a restoring force making the force negative. The preload angle (θ) is defined by the counter-clockwise rotation from the outer fixing point of the spring and a stationary airfoil at a zero angle of incidence. The angle of incidence (α) can be subtracted as it is defined at the same rotational direction. Thence, the moment caused from the springs torque is defined in Equation (3). For the reversed scenario, however, a negative spring preload angle is applied in order to fulfil the equilibrium in Equation (1).

$$\tau_s(\theta, \alpha) = -\,\kappa\,(\theta\ -\ \alpha) \tag{3}$$

A definition of the required aerodynamic moment for an equilibrium state at an angle of incidence of zero can be seen in Equation (4), acting as a reference for the applied spring preload carried out during this work. This method is more relevant than specifying preload angle or spring torque only, as it deducts the effect of the gravity of the wing and is therefore more relevant. It also offers a working approximation of Equation (1) for small angles of incidence. We name it the spring torque setting $\tau_{a,\alpha=0}$.

$$\tau_{a,\alpha=0}(\theta) = \kappa\,\theta - (x_{CoG} - x_{MPi})\, m_w\, g \tag{4}$$

For symmetrical airfoils, no lift is achieved at an angle of incidence of zero, therefore the required aerodynamic moment is theoretical. Despite this, the election to use zero degrees is decided upon the most suitable reference for this work, as cambered airfoils can reach negative angles of incidence. This is due to their lift force and pitching moment being present at an angle of incidence of zero degrees. Hence, the aerodynamic moment driving the rotation of the wing is described below, with particular focus on the differences between symmetrical and cambered wings.

It is well-known that a cambered wing generates a higher lift force than an equivalent symmetrical-shaped wing and its aerodynamic moment τ_a is not only defined by the lift force L but also by the pitching moment τ_{AC} concerning the aerodynamic centre. Two-dimensional thin airfoil theory is considered when determining the aerodynamic moment τ_a. The aerodynamic centre x_{AC} is assumed to be located near the quarter-cord location from the leading edge.

$$\tau_a = L\,(x_{AC} - x_{MPi}) + \tau_{AC}\ for\ i = 1, 2, 3, 4 \tag{5}$$

The lift force L is given by the general lift equation using the density of air ρ_{Air}, airflow velocity v, and area of the wing A. The lift coefficient for small angle α can be expressed theoretically by assuming a linear slope. Note, the lift coefficient C_L has an offset C_{L_0} for cambered airfoils, while C_{L_0}= 0 for symmetrical airfoils.

$$L = \frac{1}{2} C_L \, \rho_{Air} \, v^2 \, A \; where \; C_L = C_{L_0} + \frac{dC_l}{d\alpha} \, \alpha \tag{6}$$

The pitching moment for the cambered airfoil τ_{AC} is defined in Equation (7), where $C_{m,AC}$ is the pitching moment coefficient, while $\tau_{AC} = 0$ for symmetrical airfoils.

$$\tau_{AC} = \frac{1}{2} \, C_{m,AC} \, \rho_{Air} \, v^2 \, A \, c \tag{7}$$

When combining Equation (5), the general lift equation (Equation (6)) and Equation (7) for a static airfoil gives Equation (8).

$$\begin{aligned} \tau_a &= \frac{1}{2} \, C_L \, \rho_{Air} \, v^2 \, A \, (x_{AC} - x_{MPi}) + \frac{1}{2} \, C_{m,AC} \, \rho_{Air} \, v^2 \, A \, c \\ &= \frac{1}{2} \, \rho_{Air} \, v^2 \, A \, (C_L \, (x_{AC} - x_{MPi}) + C_{m,AC} \, c) \end{aligned} \tag{8}$$

Equations (2), (3), and (8) can then be substituted into Equation (1) that is rearranged to compare mechanical and aerodynamic forces, which creates the characteristic equation for the system in Equation (9), which is only valid for small angles.

$$\begin{aligned} \tau_a &= -\tau_s - \tau_g \\ \frac{1}{2} \rho_{Air} \, v^2 \, A \, (C_L \, (x_{AC} - x_{MPi}) + C_{m,AC} \, c) \\ &= \kappa \, (\theta - \alpha) - \cos(\alpha) \, (x_{CoG} - x_{MPi}) \, m_w \, g \end{aligned} \tag{9}$$

Therefore, at small angles, a trigger velocity of the freestream can be deduced as shown in Equation (10) below.

$$v \, (\alpha, \theta) = \sqrt{\frac{2 \, \kappa \, (\theta - \alpha) - cos(\alpha) \, (x_{CoG} - x_{MPi}) \, m_w \, g}{\rho_{Air} \, A \, (C_L \, (x_{AC} - x_{MPi}) + C_{m,AC} \, c)}} \tag{10}$$

Rearranging Equation (9) allows for the calculation of the spring preload angle required to achieve static equilibrium at the desired freestream velocity and angle of incidence given. This is shown in Equation (11).

$$\theta(\alpha, v) = \frac{\frac{1}{2}\rho_{Air}\, v^2\, A\,(C_L\,(x_{AC} - x_{MP}) + C_{m,AC}\, c) + cos(\alpha)\,(x_{CoG} - x_{MPi})\, m_w\, g}{\kappa} + \alpha \tag{11}$$

A simpler form of these equations using the symmetrical NACA 0012 at low angles of incidence have been validated with experimental results [43]. The experiment confirmed the results of the change in angle with windspeed to be within theoretical infinite and finite wing for small angles. More information of the experimental setup and validation can be found in Knight et al. [43].
For higher angles, thin airfoil theory is no longer valid and drag becomes a more dominant force that must be taken into account, alongside lift. Equation (5) becomes

$$\begin{aligned} \tau_a &= (L\cos(\alpha) + D\sin(\alpha))\,(x_{AC} - x_{MPi}) + \tau_{AC} \\ & for\; i = 1, 2, 3, 4 \end{aligned} \tag{12}$$

The characteristic equation then becomes

$$\begin{aligned} \frac{1}{2}\,\rho_{Air}\, v^2\, A\,((C_L\, cos(\alpha) + C_D\, sin(\alpha))\,(x_{AC} - x_{MPi}) + C_{m,AC}\, c) \\ = \kappa\,(\theta - \alpha) - \cos(\alpha)\,(x_{CoG} - x_{MPi})\, m_w g \end{aligned} \tag{13}$$

This results in the equation for the trigger velocity at all angles to be

$$v\,(\alpha, \theta) = \sqrt{\frac{2\,\kappa\,(\theta - \alpha) - cos(\alpha)\,(x_{CoG} - x_{MPi})\, m_w\, g}{\rho_{Air}\, A\,((C_L\, cos(\alpha) + C_D\, sin(\alpha))\,(x_{AC} - x_{MPi}) + C_{m,AC}\, c)}} \tag{14}$$

The spring preload angle can then be determined from

$$\theta(\alpha, v) = \frac{\frac{1}{2}\rho_{Air}\, v^2\, A\,((C_L\, cos(\alpha) + C_D\, sin(\alpha))\,(x_{AC} - x_{MP}) + C_{m,AC}\, c + cos(\alpha)(x_{CoG} - x_{MPi})\, m_w\, g}{\kappa} + \alpha \tag{15}$$

The next section gives an overview of the experimental setup, which includes the addition of the cambered wing.

2.2. Experimental Setup

The experiments were conducted in the University of Portsmouth's low-speed, open-circuit wind tunnel with a closed test section. A transparent box, shown in

Figure 2, encompasses the entire system and is placed within the wind tunnel test section.

(a) (b) (c)

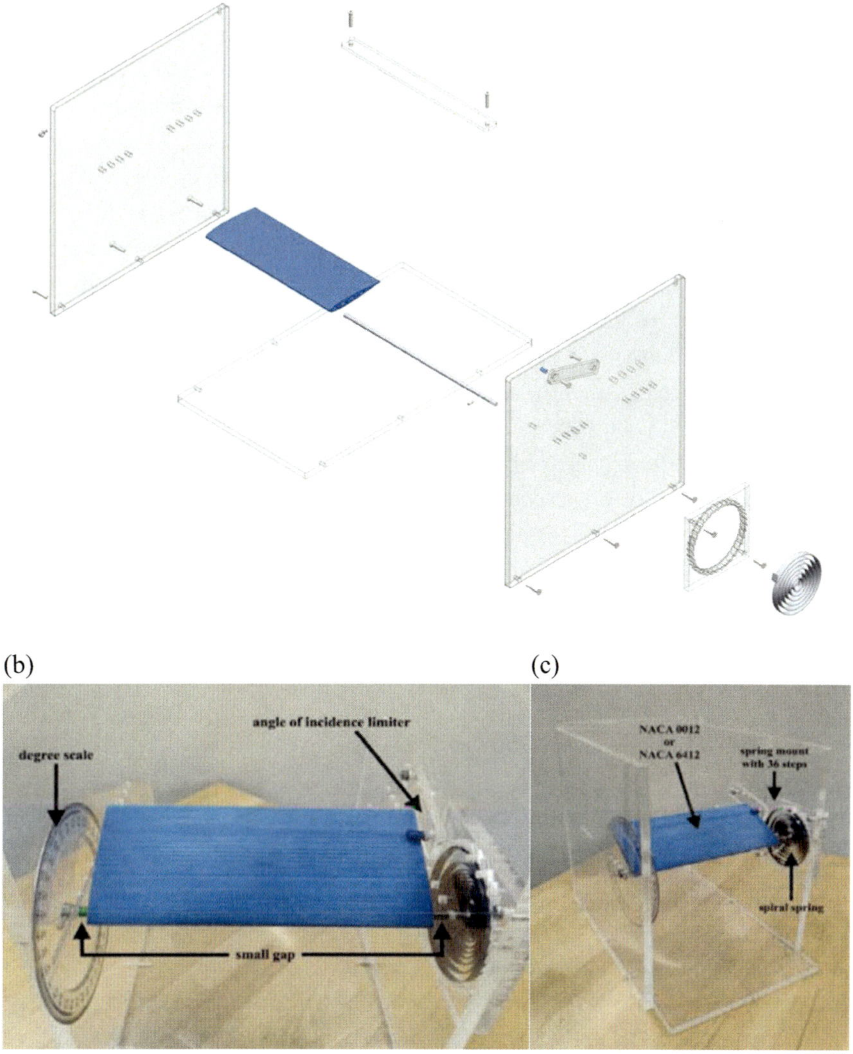

Figure 2. *Exploded view in CAD (**a**), Experimental setup (**b**) zoomed in (**c**) complete rig.*

The transparent box is made from lasered acrylic glass with dimensions of 450 mm × 262 mm × 350 mm. There are precise holes drilled into the acrylic walls used as bearings for the airfoils' mounting rod. Friction occurring in these bearings during testing is reduced by greasing them before each test. A mounting rod locks to the airfoils with a small removable pin through a recess in the side of the airfoil and hole in the mounting rod. Once fitted to the mounting rod, the airfoils have a locking pin inserted into the recess, allowing for a quick disassemble and reassemble without tools. For this mechanism, a small gap of approximately 10 mm on both sides is needed, which is shown in Figure 2a. This means the system cannot be treated completely as quasi-two-dimensional.

The spiral spring is connected to the mounting rod and a configurable mount upon the wall of the transparent box. The mounting plate has 36 notches in it, enabling the spiral spring to be positioned in 10° steps and allowing for the springs preload to be selected, as shown in Figure 2b. Preliminary tests are conducted to verify that the behaviour of the spring is linear within the required working range. The torsional spring constant is assessed to be $\kappa = 0.1643$ Nmm/deg and the spring system is calibrated by identifying the zero point, following the same procedure conducted in previous work [43]. The angle of incidence initially set for the preload angle of the spring is limited by a small pin compensating for the force produced from the spring, as shown in Figure 2a. This is referred to as the angle of incidence limiter, which can be adjusted vertically and thus allows for the initial angle of incidence to be selected. On the opposite side of the wall, a degree scale is attached for direct readings of the angle of incidence.

The system is placed in the wind tunnel and is essentially a static system in pre and post-stall regions. Above a certain flow velocity, where the aerodynamic moment overcomes the spring preload, further increases in the flow velocity will result in a change in the incidence. A constant flow velocity will result in a particular constant incidence angle, thus the system is defined as static. However, the system is dynamic in the near-stall region whereby the airfoil will oscillate at a constant flow velocity.

The blockage of the system varies from 6.3% at 0 degrees incidence to 16% at 90 degrees incidence. Throughout this paper, the angle of attack is the same as the angle of incidence. At high angles of incidence, the wake could potentially have an influence on the angle of attack due to blockage but we do not consider this effect in this paper. We acknowledge that at high angles the data should be corrected due to the wake blockage. However, for simplicity and to allow for direct comparisons, we have not corrected for wake blockage.

2.3. Airfoil Models

We use the symmetrical NACA 0012 and the cambered NACA 6412, based upon coordinate points obtained by the University of Illinois at Urbana-Champaign [46]. Both airfoils have a maximum thickness of t = 12 mm located at 30% of the chord length, wingspan of s = 228 mm, and a chord length of c = 100 mm, giving an aspect ratio of both airfoils of AR = 2.28. To ensure comparability, the properties of the two wings are as similar as possible, apart from their airfoil shape. The airfoils have four holes for the rod, which is 6 mm in diameter and used for mounting the wing. Hence, there are four pivot point locations (MP1, MP2, MP3, and MP4) located along the camber line of both airfoils, as can be seen in Figure 3. Similar weight properties for both wings were achieved by sealed cavities inside the wing, which were independently adjusted. Thus, the wings both have the same centre of gravity located at 42.4 mm (42.4% of the chord) from the leading edge and same volume despite their outer geometry being different. The airfoils are 3D printed and have refined surfaces. The density of the printing is at a 25% fill rate using the material acrylonitrile butadiene styrene (ABS), resulting in the airfoils both having a lightweight construction with a mass of 119.3 g. We do not consider surface roughness in our approach. The flow speeds investigated correspond to Reynolds numbers below 1.37×10^5 and we suspect laminar flow throughout this work [47]. At higher speeds, the surface roughness could potentially trip the flow in the boundary layer to being turbulent.

(a)

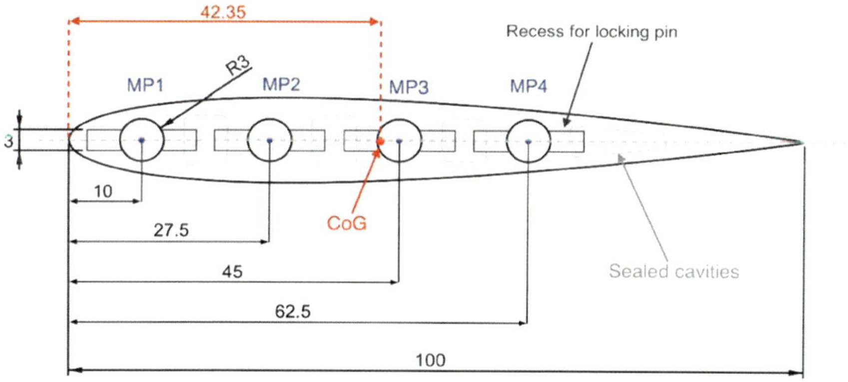

(b)

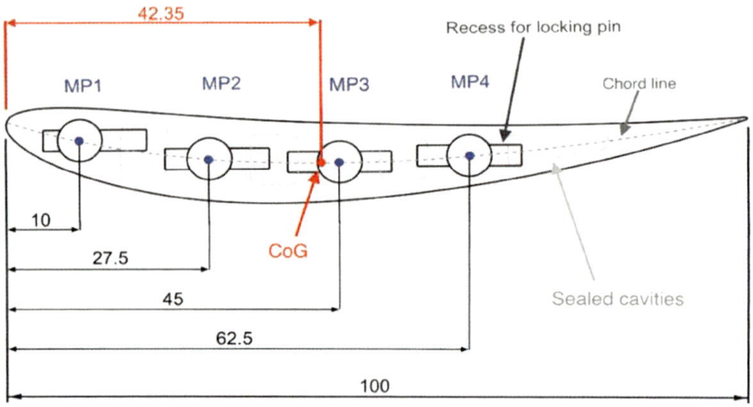

Figure 3. *Pivot point location, dimensions in mm, and the centre of gravity location (CoG) for (**a**) NACA 0012 and (**b**) NACA 6412.*

3. RESULTS AND DISCUSSION

3.1. NACA 0012 at Various Pivot Point Locations and Spring Torque Settings

The results for the response of the angle of incidence of the symmetrical airfoil while increasing the velocity of the airflow when using various spring torque settings at the pivot point locations MP1, MP2, MP3, and MP4 are presented in this section. For MP1, the spring preload angles were adjusted to match the spring torque settings in the previous study [43] as closely as possible to provide a correlation of data. The spring torque settings for MP2, MP3, and MP4 are subsequently adjusted to match as closely as possible the spring torque settings of MP1, further enabling comparisons to be made. For MP3 and MP4, as discussed before, to achieve a state of static equilibrium, the spring torque settings are inverted. However, with MP3, as the centre of gravity is very close, lower torque settings are required in order to see movement of the airfoil. MP1 and MP2 have a starting angle of incidence of 40°, which decreases as velocity increases, whereas MP3 and MP4 have a starting angle of incidence of 17°, which increases as the velocity of the flow is increases.

The results for the spring torque settings of 11.53 Nmm to 49.32 Nmm at MP1 are presented in Figure 4a. The results show the same general shape for all torque settings, shifting to a higher velocity as the springs torque increases. This is due to a larger moment required to be overcome to rotate the airfoil as the spring torque increases. However, the response of the gradient in the post-stall region (above 15°) slightly decreases as the springs torque increases, while the resistance from the spring against the force created by the airflow increases as the torque increases. In addition, the final angle of incidence at which the airfoil stopped rotating increases with higher springs' torque as the force required to rotate the airfoil is increased. As can be seen in Figure 4a, for this symmetrical profile, zero degree incidence is never achieved because at this angle the C_L is also zero and some aerodynamic force is needed to balance the spring torque. Behaviour of an oscillatory nature, which is indicated by a gap in the data, is observed at all spring torque settings, apart from the lowest tested of 11.53 Nmm. The velocity range where this is observed increases with higher spring torques. The response at pre-stall angles of incidence (below 12°) is observed as being parabolic in shape for all spring torque settings tested. The results are consistent with the previous study [43]. Results for the tests at MP2 are presented in Figure 4b, showing spring torque settings of 12.29 Nmm to 50.08 Nmm.

(a)

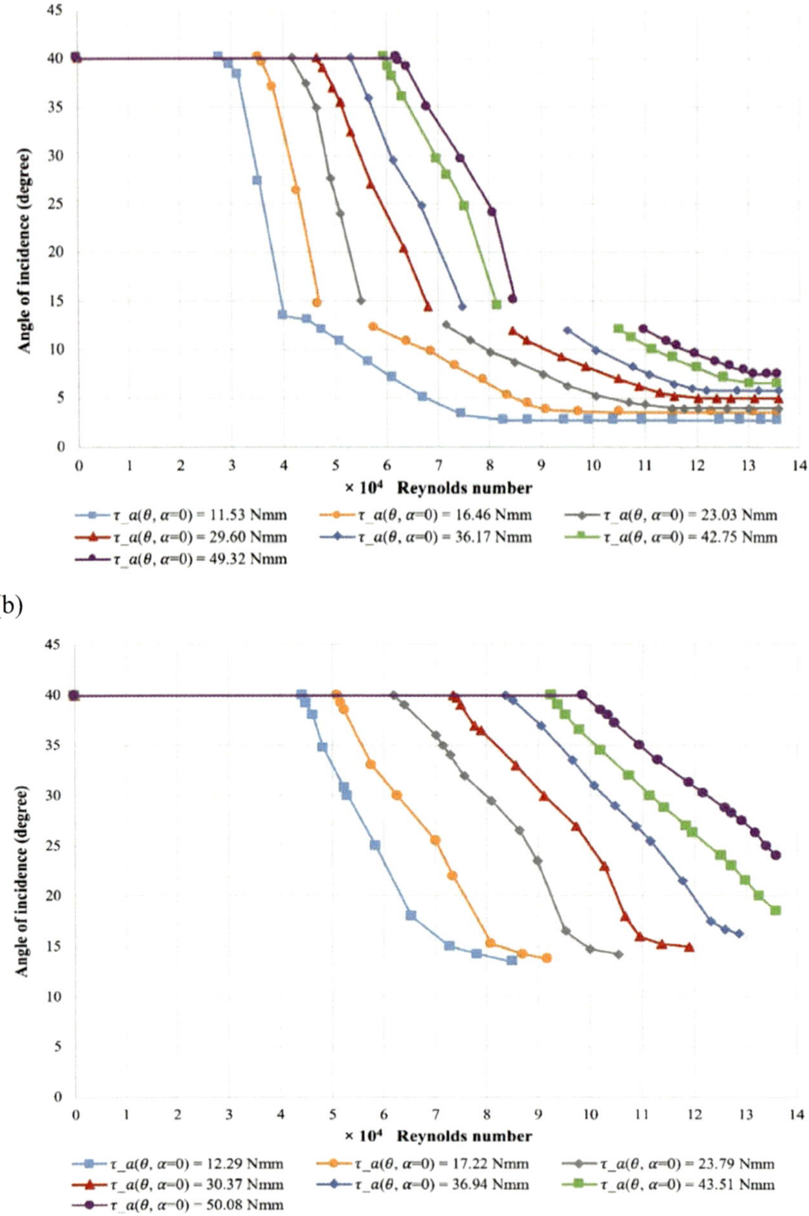

Figure 4. *Response of the angle of incidence against flow velocity for (**a**) MP1 and (**b**) MP2 of NACA 0012.*

The results also show the same general shape for all torque settings, shifting to a higher velocity as the springs torque increases due to the same reasons given for MP1. The gradient of the post-stall response also decreases as the springs torque increases, as demonstrated for MP1. With increasing spring torque, the angle of incidence where the response levels out and oscillatory behaviour is observed increases. Oscillatory behaviour is found in all torque settings, apart from 43.51 Nmm and 50.8 Nmm, which could not be studied further due to the wind speed limitations of the wind tunnel. When compared to MP1, the spring torque setting has a larger effect for MP2, as being closer to the aerodynamic centre requires more aerodynamic force to rotate the airfoil.

Results for the reversed scenario at MP3 are shown within Figure 5a with spring torque settings ranging from −1.73 Nmm to −49.38 Nmm. The results indicate the same general profile for all torque settings, shifting to a higher velocity as the springs torque increases. The response starts to level out with the parabolic curve becoming more pronounced as the spring torque increases, with the gradient decreasing at higher velocities. A greater amount of force is required to overcome the springs' moment due to the centre of the sum of the aerodynamic and gravitational moments being close to MP3. The difference between the responses is larger than with MP2, suggesting that the centre of the sum of the aerodynamic and gravitational moment is between MP2 and MP3, and closer to MP3, requiring greater aerodynamic force to rotate the airfoil. Results for various spring torque settings of −10.82 Nmm to −49.61 Nmm at MP4 are presented within Figure 5b.

(a)

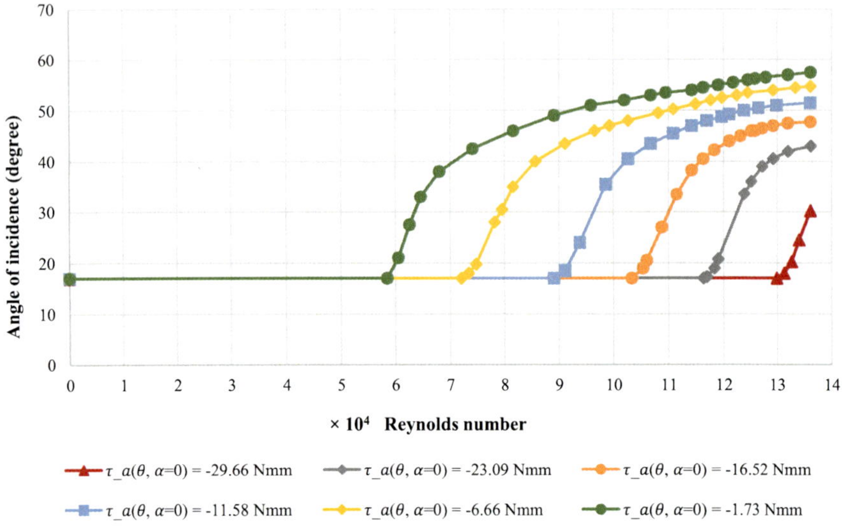

(b)

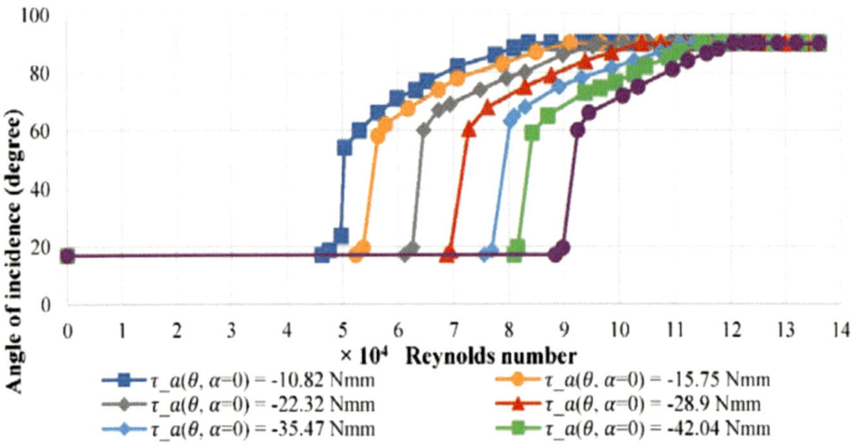

Figure 5. *Response of the angle of incidence against flow velocity for **(a)** MP3 and **(b)** MP4 of NACA 0012.*

The results show the same general profile for all torque settings, shifting to a higher velocity as the springs torque increases, as with other pivot point locations. The initial gradient for the post-stall response is large for all spring torque settings. Within this region, all torque settings show that once the angle of incidence passes

60°, the response starts to level out with lower torque settings reaching the maximum limited angle at lower velocities. The difference between responses is slightly larger than that of MP1 but lower than other pivot point locations, further proving that the centre of the sum of the aerodynamic and gravitational moment is closer to MP3.

3.2. Comparison of Pivot Point Locations for NACA 0012

The results for the response of the angle of incidence while increasing the velocity of the airflow for various pivot point locations at a similar torque setting are presented in Figure 6. The torque settings are 23.03 Nmm for MP1, 23.79 Nmm for MP2, −23.09 Nmm for MP3, and −22.32 Nmm for MP4.

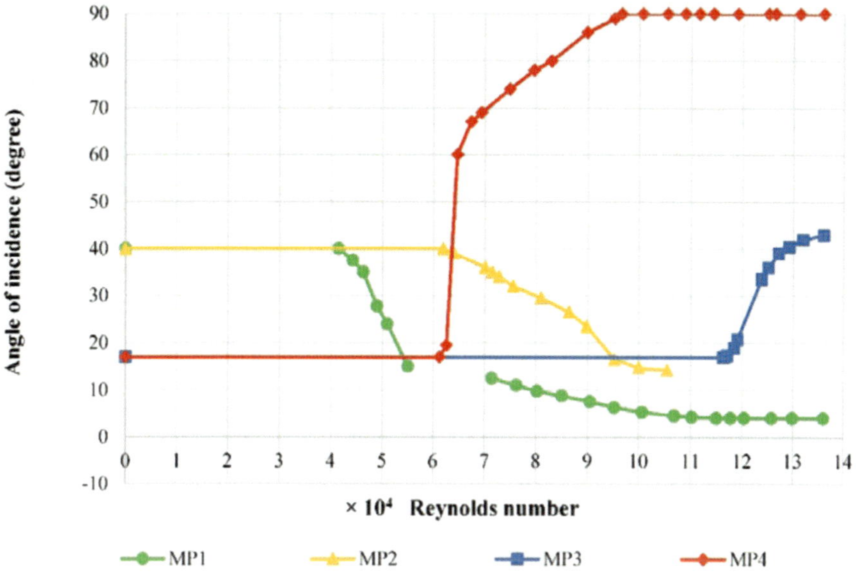

Figure 6. *Effect of pivot point location on NACA 0012.*

The results show the centre for the sum of the gravitational and aerodynamic moment for NACA 0012 is between MP2 and MP3, shown by the angle of incidence decreasing in the response to the airflow velocity for MP2 and increasing for MP3. The results show the centre is closer to MP3 than MP2, as a much increased velocity is required to produce a force to overcome the similar torques of the springs.

Comparing MP1 and MP2, the gradient of the response in the steady post-stall region (above 15°) is larger for MP1 due to MP2 being closer to the aerodynamic and centre of gravity. The aerodynamic and gravitational moments produced are smaller for MP2, requiring a higher airflow velocity to produce a force to overcome the springs' moment. In our setup, we do not reach the steady pre-stall region for MP2, which we believe will occur at speeds greater than the available maximum in our wind tunnel of 20 m/s (Re = 1.37×10^5).

The gradients for MP3 and MP4 are greater than MP1 and MP2, attributed to MP3 and MP4 having a divergent reaction as changes in aerodynamic force incur an additional aerodynamic force. MP4 has the greatest gradient due to the pivot point location being a further distance away than MP3 to the aerodynamic centre and the centre of gravity. Therefore, this requires a lower velocity to overcome the springs' moment and, once overcome, an increased divergent reaction.

For a location of the mounting axis before the centre of the sum of the aerodynamic and gravitational moments, as in MP1 and MP2, a decrease in the coefficient of lift and drag occurs, leading to a stable equilibrium if outside the stall region. For the mounting after the centre, as in MP3 and MP4, an increase in these coefficients is found, leading to a more unstable equilibrium due to divergence.

3.3. NACA 6412 at Various Pivot Point Locations and Spring Torque Settings

The results for the response of the angle of incidence while increasing the velocity of the airflow for various spring torque settings at MP1, MP2, MP3, and MP4 for the cambered NACA 6412 are presented in this section. The torque settings used are the same as for NACA 0012 to compare data. Similarly, the starting angle of incidence for MP1 and MP2 is 40°. However, a larger starting angle of incidence of 27° is used with NACA 6412 for MP3 and MP4 to avoid the unsteady stall region, which is larger for the cambered profile.

Results for various spring torque settings of 11.53 Nmm to 49.32 Nmm at MP1 are shown in Figure 7a. The results show the same general profile for all spring torque settings tested, shifting to a higher velocity as the springs' torque value increases due to the same reasons as for NACA 0012. The gradient of the post-stall response (above 25°) appears to be constant as the springs' torque increases, unlike that found with the symmetrical airfoil. In addition, the response for the lowest spring setting of 16.46 Nmm became unsteady past −2.5°. However, the behaviour once again became stable past the velocity of 16.6 m/s (Re = 1.13×10^5) at −5°. The reason is not known but thought to be due to flutter or a separation bubble forming on the top surface, which could detach and reattach in cycles [21,45,48]. Results for various spring torque settings of 12.29 Nmm to 50.08 Nmm at MP2 are presented within Figure 7b.

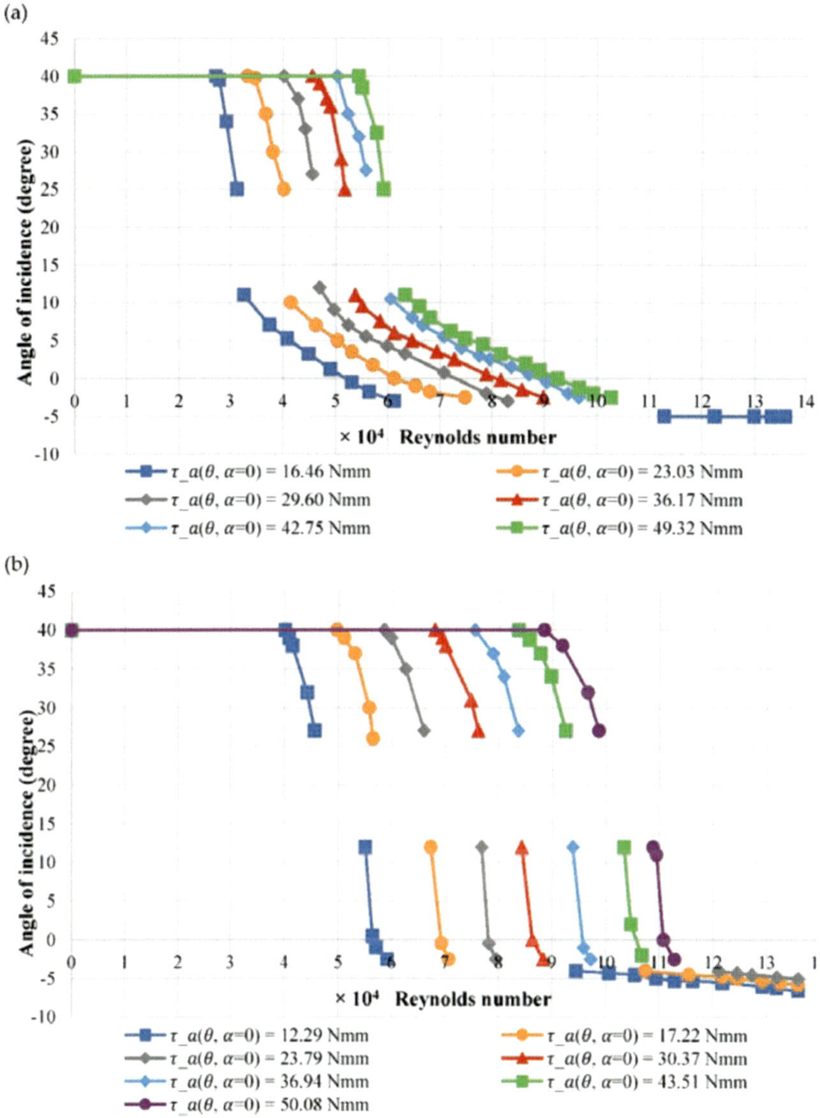

***Figure** 7. Response of the angle of incidence against flow velocity for (**a**) MP1 and (**b**) MP2 of NACA 6412.*

The results show the same general profile for all spring torque settings tested, shifting to a higher velocity as the springs' torque value increases due to the same reasons as for MP1. Here, the gradient of the post-stall response (above 27°) decreases slightly as the springs' torque increases for the same reasons as reported before for the symmetrical airfoil. The response for the lowest three spring settings of 16.46 Nmm, 17.22 Nmm, and 23.79 Nmm below −2.5° shows that the behaviour became stable past an angle of incidence of −4°, which is believed to be for the same reasons as MP1 where a separation bubble is suspected.

Results for various spring torque settings of −1.73 Nmm to −49.38 Nmm at MP3 are presented in Figure 8a. The results show the same general profile for all torque settings, shifting to a higher velocity as the springs' torque increases due to the same reasons as MP1. The initial gradient for the post-stall response, similar to MP1, appears the same as the springs' torque increases for the same reasons as MP1. However, at higher speeds, the gradient can be seen to reduce. Once torque settings past −16.52 Nmm, no movement was recorded. This is because the centre of the sum of the aerodynamic and gravitational moment is too close to MP3, thus a larger force is required to overcome the springs' moment, which would require a larger velocity than was available in the wind tunnel used. The difference between the responses is larger than with MP2, suggesting the aerodynamic centre is between MP2 and MP3, and closer to MP3, requiring greater aerodynamic force to rotate the airfoil. Results for various spring torque settings of −10.82 Nmm to −49.61 Nmm at MP4 are presented within Figure 8b.

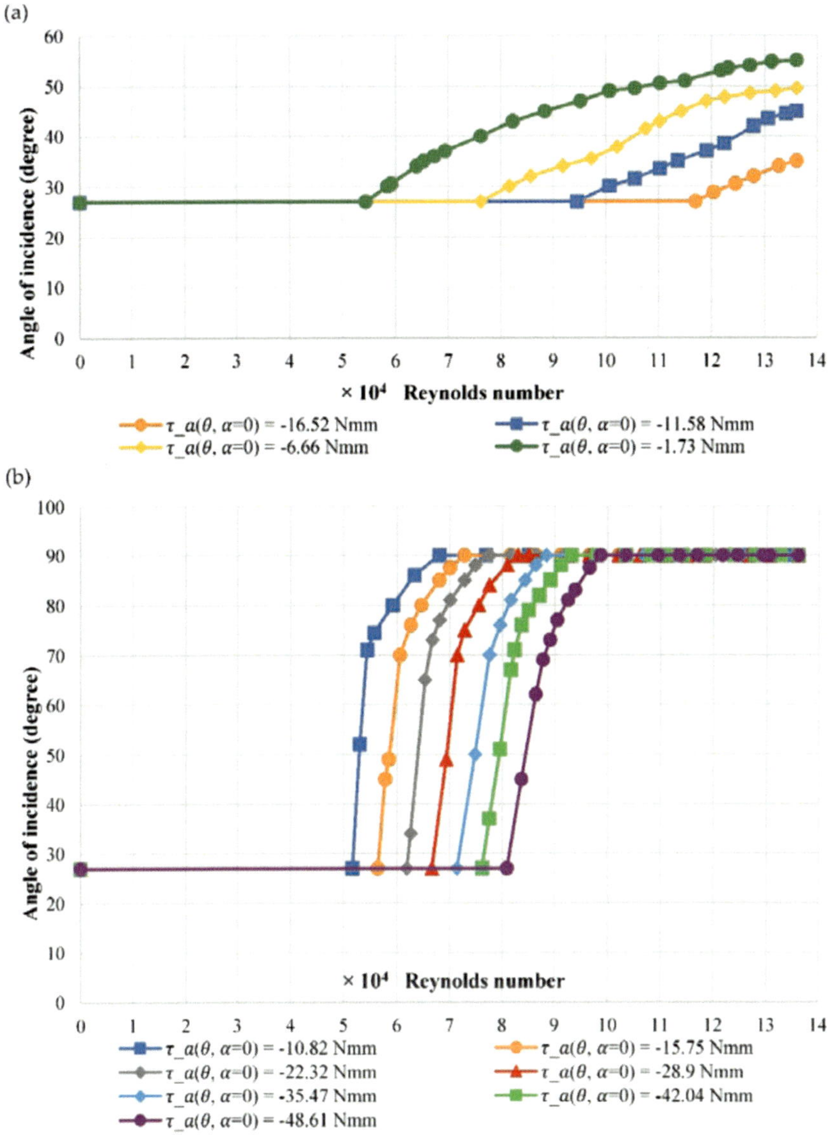

Figure 8. *Response of the angle of incidence against flow velocity for **(a)** MP3 and **(b)** MP4 of NACA 6412.*

The results show the same general profile for all torque settings, shifting to a higher velocity as the springs' torque increases due to the same reasons as for MP1. The initial gradient for the post-stall response, similar to MP2, decreases as the springs' torque increases for the same reasons. Within this range, all torque settings show that once the angle of incidence passes 70°, the responses start to level out. Using lower torque settings, the maximum limited angle is reached at lower velocities.

3.4. Comparison of Pivot Point Locations for NACA 6412

The results for the response of the angle of incidence while increasing the velocity of the airflow for the relative pivot point locations at a similar torque setting for NACA 6412 are presented in Figure 9. The spring torque settings are 23.03 Nmm for MP1, 23.79 Nmm for MP2, −23.09 Nmm for MP3, and −22.32 Nmm for MP4.

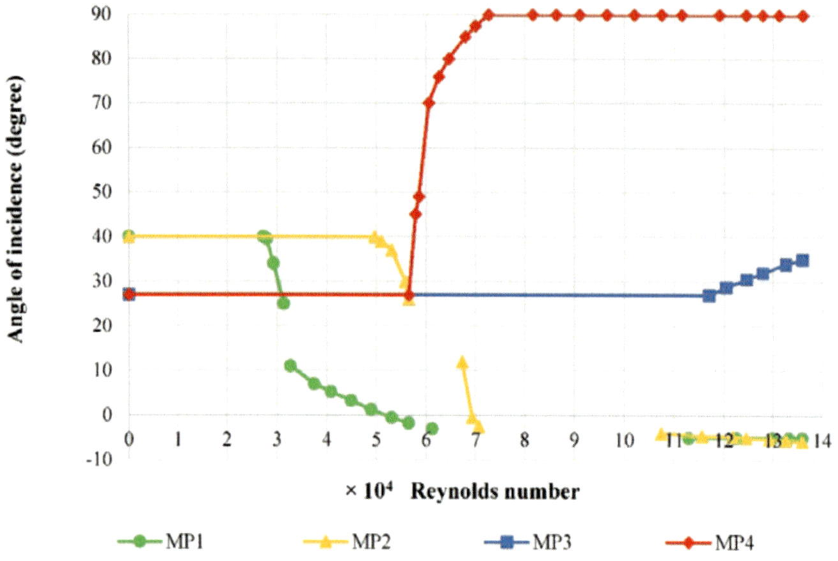

Figure 9. *Effect of Pivot point location on NACA 6412.*

The results show the centre for the sum of the gravitational and aerodynamic moment for NACA 6412 is between MP2 and MP3, indicated by the angle of incidence decreasing in the response to the airflow velocity for MP2 and increasing for MP3. The results show that the centre is closer to MP3 than MP2, shown by the velocity of the airflow being required to produce a force to overcome the springs' torque, notably higher with MP3 using similar values of spring torque.

Comparing MP1 and MP2 in Figure 9, the gradient in the steady post-stall region (40 to 27°) for MP2 is less than that of MP1, which is due to MP2 being closer to the aerodynamic and gravitational centre. The aerodynamic and gravitational moments produced are smaller, requiring a higher airflow velocity to produce a force to overcome the springs' moment. Conversely, within the steady pre-stall region (12 to −2°), the gradient for MP2 is shown in Figure 9 to be much larger than MP1. The reason for this is believed to be caused by the camber of the lower surface protruding down in front of MP2, creating a rapid change from suction to high pressure on that lower surface over a small change in the freestream velocity. This produces a moment that changes the sign, adding to that produced in the same rotational direction behind the pivot point location on the upper surface, resulting in the airfoils nose rapidly moving upwards. At MP1, this protruding area is behind the pivot point location, acting as a counteracting force, thus the change is more gradual.

MP3 has a lower gradient compared to other pivot point locations primarily due to MP3 being located close to the aerodynamic centre. The suction from the protruding area on the lower surface counteracts the high pressure from the upper surface close to the trailing edge, which rotates the airfoil upwards. At MP4, the gradient is steeper as the aerodynamic centre is further away; hence, small changes to the aerodynamic force have a larger effect on the moment produced, leading to even higher incidences and divergence.

For a location of the mounting axis before the aerodynamic centre (MP1 and MP2), a decrease in the incidence and therefore coefficient of lift and drag occurs. For the mounting after the aerodynamic centre (MP3 and MP4), an increase in the incidence and therefore an increase in these coefficients is observed. In the post-stall region, above 27°, the gradients and hence the change in incidence is larger the further the pivot point location is from the aerodynamic centre. Conversely, the change is less rapid in the pre-stall region for angles between 12° and −2° for the airfoils tested at MP1 and MP2.

3.5. The Comparison between Cambered and Symmetrical Airfoils

The figures in this section show the response of the angle of incidence while increasing the velocity of the airflow for each of the relative pivot point locations at the same torque settings. This enables a comparison of the differences in airfoils at each of the pivot point locations. Figure 10a shows the effect of the airfoil shape for MP1 at a spring toque setting of 36.17 Nmm. For the NACA 6412 cambered airfoil, a lower airflow velocity is required to initiate movement, with it reaching lower angles of incidence than the symmetrical airfoil. This is caused by an aerodynamically improved performance of NACA 6412, with the airfoil producing a larger amount of downforce than NACA 0012. The response gradient for NACA 6412 can be seen in Figure 9a, slightly larger in both the post-stall region and pre-stall region. The unsteady near-stall region for NACA 6412 occurs from 25° to 12°, whereas for NACA 0012, it occurs from 15° to 12°. The larger region for NACA 6412 is due to the airfoils cambered shape, resulting in a delayed and more gradual stall, which results in higher fluctuations of pressure and hence unsteadiness. Figure 10b shows the effect of airfoil shape for MP2 at a spring toque setting of 30.37 Nmm.

(a)

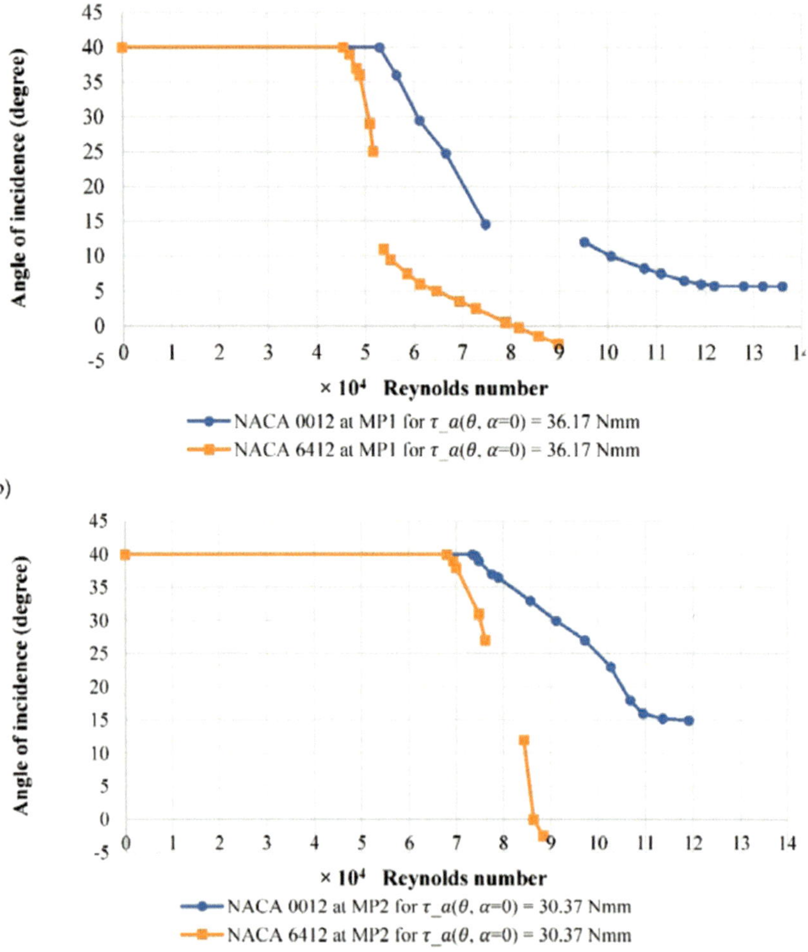

Figure 10. *Effect of airfoil shape at (**a**) MP1 and (**b**) MP2.*

Once again for NACA 6412, a lower airflow velocity is required to initiate movement, with it reaching lower angles of incidence into the steady pre-stall region, similar to that found with MP1. Comparing the steady post-stall region, similarly, the gradient is larger for NACA 6412 and the unsteady near-stall region is the same as MP1. Conversely, the gradient for NACA 6412 in the pre-stall region with MP2 is larger than that for MP1, which, as previously mentioned, is suspected to be due to a rapid change in moment. The pre-stall region is not reached with NACA 0012 using MP2 and would require larger airflow velocities than were available.

Figure 11a shows the effect of airfoil shape for MP3 at a spring toque setting of −6.66 Nmm. The starting angles are different when comparing airfoils at MP3. Nevertheless, at 27°, a slightly smaller velocity is required to initiate movement of NACA 6412; however, it has a decreased gradient, reaching a lower angle of incidence at higher wind speeds. This is partly due to NACA 6412 producing a slightly larger aerodynamic force, initiating movement at a slightly lower velocity. The gradient is slightly less when compared to the NACA 0012 but similar at higher velocities. The cambered airfoil produces a larger amount of downforce at a given incidence due to the increased turning of the airflow. Figure 11b shows the effect of airfoil shape for MP4 at −42.04 Nmm.

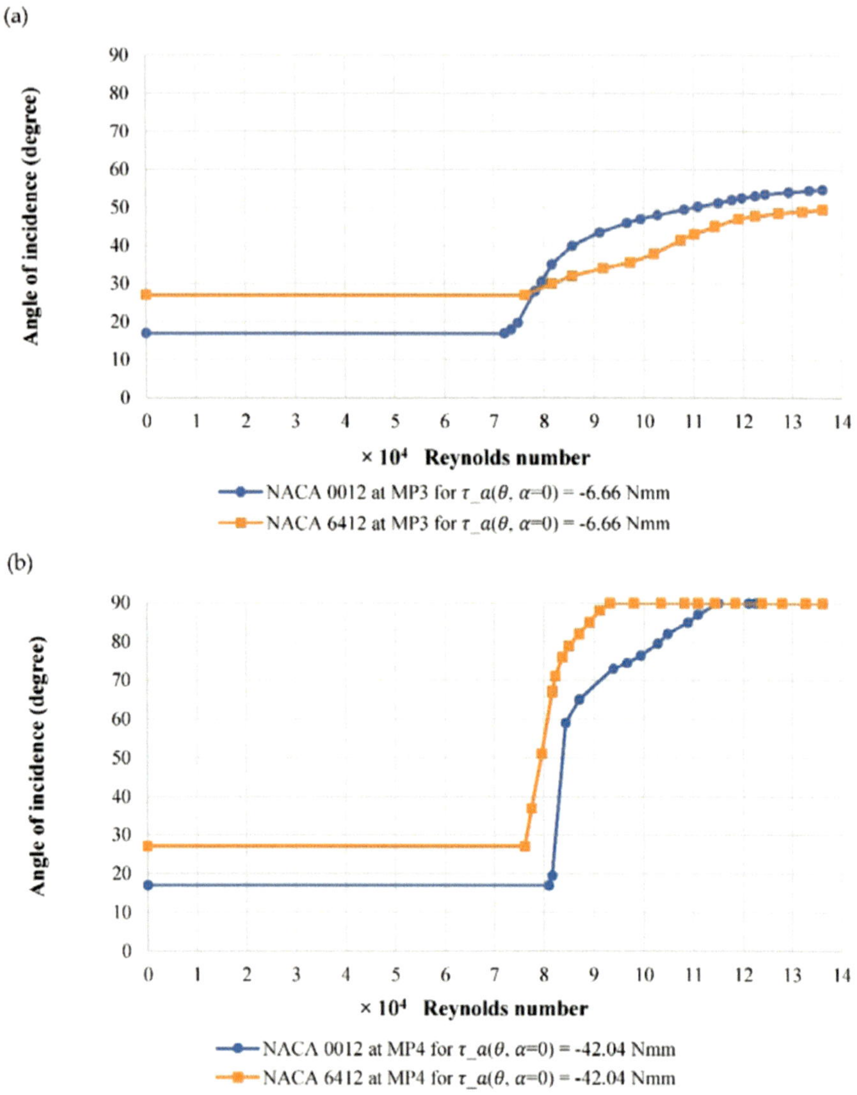

Figure 11. *Effect of airfoil shape at (**a**) MP3 and (**b**) MP4.*

At 27°, a smaller velocity is required to initiate movement of NACA 6412 at MP4 when compared to MP3. Conversely, the NACA 6412 reaches the maximum limited angle of 90° at a lower velocity when compared to the symmetrical airfoil, unlike MP3. The gradients are similar up to 60° and thereafter both gradients reduce but the symmetrical airfoil can be seen to reduce more, reaching the limiting angle at a higher wind speed.

3.6. Unsteady Aeroelastic Behavior

The unsteady behaviour of aerodynamic devices is a detrimental phenomenon in most applications, for example in race car aerodynamics. Oscillation about the axis of rotation is a potential threat to the applicability of a spring based aeroelastic system. Indeed, self-sustaining oscillation is observed at all torque settings except $\tau_{a,\alpha=0}$ = 10.71 Nmm, the lowest setting in this experiment. Therefore, this behaviour will be discussed in more detail in the following, where for simplicity the experiments are limited to the symmetrical wing NACA 0012.

The oscillation behaviour of the system is examined at selected spring preload angles from θ = 336.7° to θ = 536.7° or $\tau_{a,\alpha=0}$ = 17.26 Nmm to $\tau_{a,\alpha=0}$ = 50.02 Nmm. The vibration only occurs at an angle of attack in the range of 12° to 15°. This is observed equally for all torque settings and corresponding velocities. The Reynolds numbers of the investigated measuring points range from 4.6×10^4 to 1.1×10^5.

To explain this phenomenon, the stall behaviour of the NACA 0012 wing, which depends on the Reynolds number, must be taken into account. It is known that the airflow separates from the wing at an angle of attack greater than the stall angle. The observed oscillation occurs at an angle of attack slightly above the expected stall angle. However, compared to the stall behaviour of a fixed wing, the present system is more complex.

It is believed that the aeroelastic phenomenon of dynamic stall is recognised. Dynamic stall is generally a transient aerodynamic effect in which the airflow alternates between attachment and detachment. This hypothesis is strongly suggested by comparison with other published literature [44, 49, 50]. Although the existence of an laminar separation bubble (LSB) cannot be proven with the current experimental setup, it is known that detachment and reattachment of an LSB could be the cause of oscillation [51, 52]. Indeed, oscillation is not observed at incidence angles above 15°.

The absence of oscillation at the lowest spring preload ($\tau_{a,\alpha=0}$ = 10.71 Nmm), could possibly indicate detachment of the airflow without laminar reattachment to the wing. The Reynolds number at the critical angle of incidence is 3×10^3 to

3.7×10^3, which is below the oscillation limit proposed by Poirel et al [42]. However, it has to be mentioned that the absence of oscillations could also be caused by the friction of the bearings not being overcome at this low airflow velocity. The oscillation is investigated in more detail by using high-speed video images. It is shown that the frequency of the symmetrical blade is constant at a fixed flow velocity. The maximum deviation is less than 3 % from the average of all measured frequencies within the measurement time span of 15 to 45 seconds. The reduced frequency K is defined as

$$K = \frac{f\, c\, \pi}{v} \tag{16}$$

where f is the measured frequency. This dimensionless frequency is used to provide a measure of discontinuity and to allow comparison of the oscillation with results from other experiments with different dimensions and airflow velocities. The torque setting $\tau_{a,\alpha=0}$ is used to indicate the gradual increase in spring preload and plotted with the corresponding Reynolds number to show the interaction of these three parameters, which is shown in Figure 12.

(a)

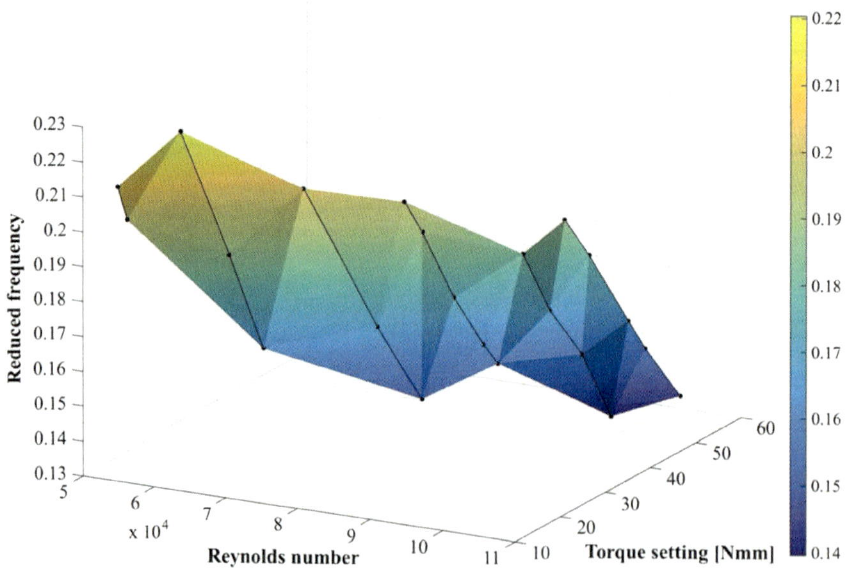

(b)

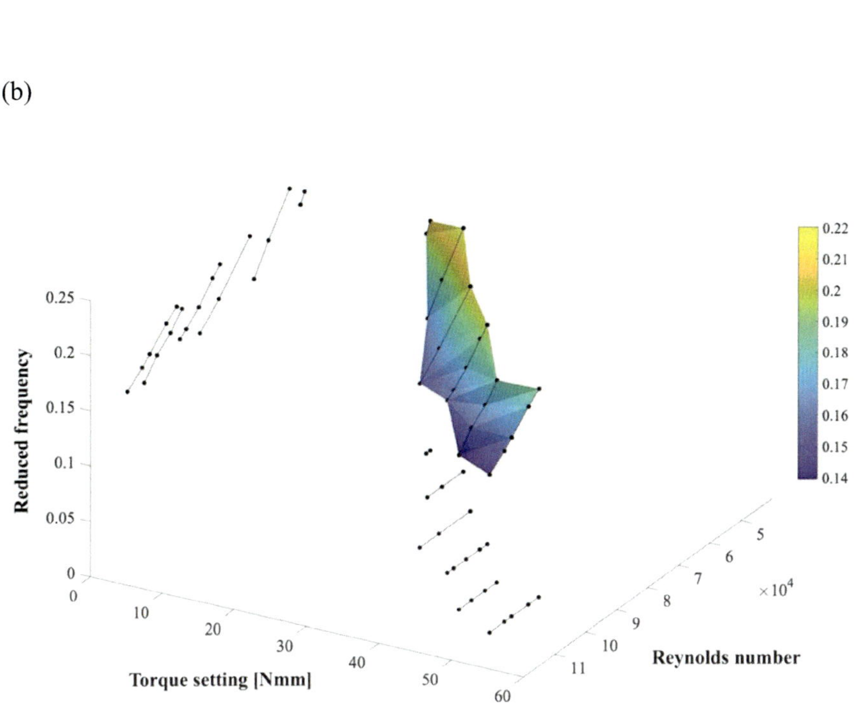

Figure 12. *Illustration of unsteadiness region with reduced frequency values of the NACA 0012 wing at various Reynolds numbers and spring preloads given as torque setting.*

The results reveal that within the tested range, the peak of discontinuity is at $\tau_{a,\alpha=0}$ = 23.82 Nmm and a Reynolds number of 5.51×10^4. A decreasing trend of reduced frequency is observed both at higher Reynolds numbers and at high spring preload. Reduced frequencies are observed at a minimum of 0.14 and a maximum of 0.22. All the reduced frequencies observed are very high, so the oscillation is generally considered to be highly unsteady. Similar reduced frequencies were observed by Poirel et al [42] in the range of 0.1 to 0.2 at Reynolds numbers in the range of 4.5×10^4 up to 1.3×10^5. The simple solution to avoid the oscillations is to avoid angles of incidence greater than 12°. In the experimental setup used here, the angle of incidence limiter can be set to 12° to avoid any unsteady behaviour. [43]

4. CONCLUSIONS

From the results obtained, the study has proven the concept of an airfoil passively adjusting to a freestream velocity. The adjustment is highly dependent on pivot point location and spring preload, resulting in either a reduction or increase in both downforce and drag. From the results, it is concluded that pivot point location plays the largest role in whether the lift and drag coefficients increase or decrease and when concerning the rate of change when subjected to a freestream velocity. When the sum of the aerodynamic and gravitational moment was negative (leading edge down, MP3 and MP4), an increase in lift and drag coefficients was observed due to the increase in incidence. Conversely, when the aerodynamic and gravitational moment sum was positive (leading edge up, MP1 and MP2), a decrease in lift and drag coefficients occurred due to the decrease in incidence. The centre of the sum of the aerodynamic and gravitational moments was determined to be between MP2 and MP3 for both airfoils, positioned slightly closer to MP3. When the aerodynamic centre and centre of gravity was closer to the mounting axis, this produced smaller moments, decreasing the rate of change. Where the pivot point location was closer to the centre of the sum of the aerodynamic and gravitational moments, the velocity range in which unsteady behaviour occurred increased.

The results also showed that spring torque played a role in the initiation, rate of change, and final angle of incidence achieved. As the springs' torque increased, the wind speed required to change the angle of incidence increased. In addition, with higher windspeeds, the rate of change in the angle of incidence decreased. However, as the springs' torque increased, the velocity range in which unsteady behaviour occurred also increased. The cambered NACA 6412 produced a larger aerodynamic force and therefore aerodynamic moment when compared with NACA 0012 airfoil at a given windspeed. This larger force enabled changes to the angle of incidence at lower velocities. However, the cambered NACA 6412 was also subject to increased velocity ranges and angles of incidence of unsteady behaviour.

The springed wing system described here could represent a real wing with support structure that is elastic. The pivot point and spring torque setting represent the mounting location and strength of support, respectively. The symmetrical or cambered wing could be used in the application of reducing the drag of road vehicles at higher windspeeds. To avoid unsteady fluctuations in forces, it would be advised to use an initial setting no higher than 12°. The symmetrical airfoil has a larger speed range and no upper limit to its use. The cambered profile is more efficient, although a final setting no larger than −2° should also be used to avoid unsteady fluctuations. The value of the spring preload can be selected according to the desired velocity at which the road vehicle is intended to have a reduced drag and the pivot point location can be adjusted according to the intended rate of drag reduction with velocity. For the opposite application of increased stability with velocity, the cambered profile demonstrated a larger velocity range and lower rate of change when compared to the symmetrical profile. Similarly, the initial velocity can be triggered by selecting a spring preload. The use of symmetrical or cambered profiles with pivot point locations ahead of the centre of pressure could be used with hydrofoils and rigid sails to prevent damage at higher than intended wind velocities, although the effects of heave may have to be considered in these applications alongside aerospace applications. Conversely, they could also be used in shielding fences to block flow at higher wind speeds, provided that a suitable support structure is used. In the same manor, the trigger velocities can be selected with the preload and rate of change with the position of the pivot point.

It was found that the oscillation of the symmetric wing is periodic and occurs in the region just above the typical stall angle. Although not directly proven, there are strong arguments that the oscillation is caused by dynamic stall. This unsteady behaviour is considered a serious concern for an automotive or motorsport application. The behaviour is considered to be highly unsteady within the observed unsteady region of Reynolds numbers 4.6×10^4 up to 1.1×10^5 and peaks at a Reynolds number of 5.51×10^4. In general, operation is only recommended at an angle of attack before stall that includes a safety margin to achieve consistent, stable results.

The experimental system's setup had some notable challenges and limitations. At low values of spring torque settings, it is believed the results could be affected by friction in the bearings, which could have produced slightly distorted results. Steady pre-stall measurement is also attempted at MP3 and MP4 but are not reported here due to the high sensitivity and difficulties in achieving reliable results.

Recommendations for future work include an investigation of the effects at higher Reynolds numbers, which would require more advanced wind tunnel facilities capable of higher airflow velocities. Another area of interest would be the steady pre-stall region for MP3 and MP4, for which a wind tunnel with smaller incremental changes in velocity to capture the effects is needed. Finally, further investigation could be carried out into how this type of fluid–structure interaction can be achieved by the use of anisotropic materials, such as carbon fibre, and how the layup of the material can change the behaviour of the response. The work reported here is performed with torsional springs obeying Hooke's law. Further work is proposed to study spring-mounted wings with variable spring constants dependent on deformation, as well as wings with elastic support structures of various strengths and configurations. In addition to this, the results can be used as an initial benchmark to validate numerical simulations before tackling more complex fluid–structure interactions.

Funding: This research study received no external funding.

Data Availability Statement: Data supporting reported results can be requested from corresponding author.

Acknowledgments: The authors would like to acknowledge the technical support provided by Robert Elliot at the University of Portsmouth.

Conflicts of Interest: The authors declare no conflicts of interest.

REFERENCES

1. De Borst, R.; Nithiarasu, P.; Tezduyar, T. *Computational Fluid Structure Interaction*, 1st ed.; John Wiley & Sons Ltd.: Chichester, UK, 2013.
2. Dowell, E.H.; Clark, R.; Peters, D. *A Modern Course in Aeroelasticity*, 4th ed.; Kluwer Academic Publishers: New York, NY, USA, 2004.
3. Rathore, J.; Dhingra, K.; Malik, R. Fluid structure interaction analysis of tapered wing of aircraft. *Int. J. Eng. Res. Technol.* **2019**, *8*, 307–319.
4. Wang, W.; Yan, Y. Strongly coupling of partitioned fluid-solid interaction solvers using reduced-order models. *Appl. Math. Model.* **2010**, *34*, 3817–3830.
5. Rose, B.R.; Junu, G.R.; Manivel, M. Partly coupled fluid structure interaction analysis of an aircraft wing at subsonic speeds. *Int. J. Mech. Mechatron. Eng.* **2014**, *14*, 22–29.
6. Fung, Y.C. *An Introduction to the Theory of Aeroelasticity*; University of California: Los Angeles, CA, USA; Dover Publications, Inc.: San Diego, CA, USA, 1993.
7. Nguyen, N.; Tuzcu, I. Aeroelastic Flight Dynamics of Flexible Aircraft with Inertial and Propulsive Force Interactions. In *AIAA Atmospheric Flight Mechanics Conference*; American Institute of Aeronautics and Astronautics: Chicago, IL, USA, 2009; pp. 1–23.
8. Collar, A. The First Fifty Years of Aeroelasticity. *Aerospace* **1978**, *5*, 12–20.
9. Liauzun, C.; Piet-Lahanier, N. Aeroelasticity and Structural Dynamics. *Aerosp. J.* **2018**, *14*, 1–3.
10. Chinmaya, P.; Venkatasubramani, S. Aeroelasticity—In general and flutter phenomenon. In Proceedings of the 2009 Second International Conference on Emerging Trends in Engineering & Technology, Nagpur, India, 16–18 December 2009; pp. 81–85.
11. Sadraey, M. *Aircraft Design, A Systems Engineering Approach*, 1st ed.; John Wiley & Sons, Ltd.: Chichester, UK, 2013.
12. Afonso, F.; Leal, G.; Vale, J.; Oliveira, É.; Lau, F.; Suleman, A. Linear vs non-linear aeroelastic analysis of high-aspect-ratio wings: Effect of chord and taper-ratio variation. In *Congress on Numerical Methods in Engineering*; APMTAC, Lisbon, Portugal, 2015; pp. 1–9.
13. Yurkovich, R. Status of unsteady aerodynamic prediction for flutter of high-performance aircraft. *J. Aircr.* **2003**, *40*, 832–842.

14. Rose, B.R.; Jinu, G.R. A Study on Aeroelastic Flutter Suppression and Its Control Measures—Past and Future. *Int. J. Eng. Technol.* **2014**, *6*, 960–973.
15. Wang, Z.; Zhang, W.; Wu, X.; Chen, K. A novel unsteady aerodynamic Reduced-Order Modelling method for transonic aeroelastic optimization. *J. Fluids Struct.* **2018**, *82*, 308–328.
16. Giovanneti, L. Fluid Structure Interaction Testing, Modelling and Development of Passive Adaptive Composite Foils. Ph.D. Thesis, University of Southampton, Southampton, UK, 2017.
17. Bose, C.; Badrinath, S.; Gupta, S. Dynamical stability analysis of a fluid structure interaction system using a high fidelity Navier-Stokes solver. *Procedia Eng.* **2016**, *144*, 883–890, doi:10.1016/j.proeng.2016.05.107.
18. Stanciu, V.; Stroe, G.; Andrei, I. Linear models and calculations of aeroelestic flutter. *U.P.B. Sci. Bull. Series D* **2012**, *74 (2)*. https://www.scientificbulletin.upb.ro/rev_docs_arhiva/fullc07_145478.pdf (accessed on 26/8/21).
19. Sidlof, P.; Vlček, V.; Štěpán, M. Experimental investigation of flow-induced vibration of a pitch—Plunge NACA 0015 airfoil under deep dynamic stall. *J. Fluids Struct.* **2016**, *67*, 48–59.
20. Rivera, J.; Dansberry, B.; Bennett, R.; Durham, M.; Silva, W. NACA 0012 benchmark model experimental flutter results with unsteady pressure distributions. In Proceedings of the 33rd Structures, Structural Dynamics and Materials Conference, Dallas, TX, USA, 13 April 1992.
21. Khatir, T.; Djillali, B.; Khatir, S.; Abdelkader, L.; Wahab, M.A. Experimental and numerical investigation of flutter phenomenon of an aircraft wing (NACA 0012). *Mechanika* **2017**, *23*, 562–566.
22. Goyaniuk, L.; Poirel, D.; Benaissa, A. Pitch–Heave Symmetric Stall Flutter of a NACA 0012 at Transitional Reynolds Numbers. *AIAA J.* **2020**, *58*, 3286–3298.
23. Thill, C.; Etches, J.; Bond, I.; Potter, K.; Weaver, P. Morphing skins. *Aeronaut. J.* **2008**, *112*, 117–139.
24. Gaspari, A.; Moens, F. Aerodynamic shape design and validation of an advanced high-lift device for a regional aircraft with morphing droop nose. *Int. J. Aerosp. Eng.* **2019**, *2019*, 1–21.
25. Basaeri, H.; Yousefi-Koma, A.; Zakerzadeh, M.R.; Mohtasebi, S.S. Experimental study of a bio-inspired robotic morphing wing mechanism actuated by shape memory alloy wires. *Mechatronics* **2014**, *24*, 1231–1241.

26. Barbarino, S.; Bilgen, O.; Ajaj, R.; Friswell, M.I.; Inman, D.J. A review of morphing aircraft. *J. Intell. Mater. Syst. Struct.* **2011**, *22*, 823–877.
27. Su, W.; Swei, S.; Zhu, G. Optimum wing shape determination of highly flexible morphing aircraft for improved flight performance. *J. Aircr.* **2016**, *53*, 1305–1316.
28. Barnard, R.H. *Road Vehicle Aerodynamics*, 3rd ed.; Mechaero Publishing, St. Albans, UK, 2010.
29. McBeath, S. *Competition Car Aerodynamics*, 3rd ed.; Veloce Publishing Ltd.: Poundbury, UK; Boston, MA, USA; Dorset, UK, 2017.
30. Thuwis, G.A.; De Breuker, R.; Abdalla, M.M.; Gürdal, Z. Aeroelastic tailoring using lamination parameters—Drag reduction of a Formula One rear wing. *Struct. Multidiscip. Optim.* **2010**, *41*, 637–646.
31. Kajiwara, S. Passive variable rear-wing aerodynamics of an open-wheel racing car. *Automot. Engine Technol.* **2017**, *2*, 107–117, doi:10.1007/s41104-017-0021-9.
32. Wolf, T. The aerodynamic development of the new Porsche Cayenne. *Proc. Inst. Mech. Eng. Part D J. Automob. Eng.* **2020**, *234*, 390–408.
33. Knight, J.; McConnell, M.R.; Ledger, A.; Azcueta, R. Fluid structure interaction of hydrofoils. In *Proceedings of the ICMET Oman 2019—International Conference on Marine Engineering and Technology, Military Technological College Muscat, Muscat, Oman, 5–7 November 2019*; Zenodo: Muscat, Oman, 2019.
34. Zarruk, G.A.; Brandner, P.; Pearce, B.W.; Phillips, A.W. Experimental study of the steady fluid—Structure interaction of flexible hydrofoils. *J. Fluids Struct.* **2014**, *51*, 326–343.
35. Scott, S.; Capuzzi, M.; Langston, D.; Bossanyi, E.; McCann, G.; Weaver, P.; Pirrera, A. Effects of aeroelastic tailoring on performance characteristics of wind turbine systems. *Renew. Energy* **2017**, *114*, 887–903, doi:10.1016/j.renene.2017.06.048.
36. Abbaspour, M.; Radmanesh, A.R.; Soltani, M.R. Unsteady flow over offshore wind turbine airfoils and aerodynamic loads with computational fluid dynamic simulations. *Int. J. Environ. Sci. Technol.* **2016**, *13*, 1525–1540, doi:10.1007/s13762-016-0995-2.
37. Xu, Y.Z.; Virk, M.S.; Knight, J.; Mustafa, M.Y.; Haritos, G. Factors Influencing the Performance of Porous Wind Shields. *Appl. Mech. Mater.* **2013**, *321–324*, 799–803, doi:10.4028/www.scientific.net/amm.321-324.799.
38. Stanford, B.; Ifju, P.; Albertani, R.; Shyy, W. Fixed membrane wings for micro air vehicles: Experimental characterization, numerical modeling,

and tailoring. *Prog. Aerosp. Sci.* **2008**, *44*, 258–294, doi:10.1016/j.paerosci.2008.03.001.
39. Aboelezz, A.; Hassanalian, M.; Desoki, A.; Elhadidi, B.; El-Bayoumi, G. Design, experimental investigation, and nonlinear flight dynamics with atmospheric disturbances of a fixed-wing micro air vehicle. *Aerosp. Sci. Technol.* **2019**, *97*, 105636, doi:10.1016/j.ast.2019.105636.
40. Panta, A.; Fisher, A.; Mohamed, A.; Marino, M.; Xu, R.; Liu, H.; Watkins, S. Low Reynolds number aerodynamics of leading-edge and trailing-edge hinged control surfaces: Part I statics. *Aerosp. Sci. Technol.* **2019**, *99*, 105563, doi:10.1016/j.ast.2019.105563.
41. Cattafesta, L.; Bahr, C.; Mathew, J. Fundamentals of Wind-Tunnel Design. In *Encyclopaedia of Aerospace Engineering*, 1st ed.; Blockley, R., Shyy, W., Eds.; John Wiley & Sons Ltd.: Chichester, UK, 2010; pp. 1–10.
42. Poirel, D.; Harris, Y.; Benaissa, A. Self-sustained aeroelastic oscillations of a NACA 0012 airfoil at low-to-moderate Reynolds numbers. *J. Fluids Struct.* **2008**, *24*, 700–719, doi:10.1016/j.jfluidstructs.2007.11.005.
43. Knight, J.; Fels, S.; Haritos, G.; Carolus, T. Fluid-Structure Interaction of a Spring-Mounted Symmetrical Rigid Wing for Drag Reduction of Cars at Higher Wind Velocities. *SAE Tech. Pap.* **2020**, *1*, 5037.
44. McAllister, K.W.; Carr, L.; McCrokskey, W.J. Dynamic Stall Experiments on the NACA 0012 Airfoil. *NASA Tech. Pap.* **1978**, *1100*. https://core.ac.uk/download/pdf/42875301.pdf (accessed on 26/8/21).
45. Akbari, M.; Price, S. Simulation of dynamic stall for a NACA 0012 airfoil using a vortex method. *J. Fluids Struct.* **2003**, *17*, 855–874.
46. UIUC Applied Aerodynamics Group. UIUC Airfoil Data Site. Available online: http://m-selig.ae.illinois.edu/ads/coord/n0012.dat (accessed on 26/8/21).
47. Yousefi, K.; Razeghi, A. Determination of the Critical Reynolds Number for Flow over Symmetric NACA Airfoils. In AIAA 2018–0818. Session: Stability and Transition II: Shear Flows II. Available online: https://doi.org/10.2514/6.2018-0818 (accessed on 26/8/21).
48. Rinoie, K.; Takemura, N. Oscillating behaviour of laminar separation bubble formed on an aerofoil near stall. *Aeronaut. J.* **2004**, *108*, 153–164.
49. McCrokskey, W.J., Carr, L., and McAllister, K.W., "Dynamic Stall Experiments on Oscillation Airfoils," *AIAA J.* **1976,** *14*, 57-63.
50. Akbari, M. and Price, S., "Simulation of Dynamic Stall for a NACA 0012 Airfoil Using a Vortex Method," *Journal of Fluids and Structures*. **2003**, *17*, 855-874.

51. Wang, S., Alam, M. et al., "Effects of Reynolds Number and Turbulent Intensity on a Low Reynolds Number Airfoil," *Physics of Fluids*. **2014,** *26*, 115107, 1-25.
52. Rinoie, K. and Takemura, N., "Oscillating Behaviour of Laminar Separation Bubble Formed on an Aerofoil near Stall," *The Aeronautical J.* **2004**, *108*, 1081, 153-164.

Publisher: Eliva Press Global Ltd

Email: info@elivapress.com

Made in United States
Orlando, FL
05 November 2026

90219449R00029